P. R. BREGAZZI

A KEY TO THE BRITISH FRESHWATER

Cyclopid and Calanoid Copepods

with ecological notes

by

J. P. HARDING, Ph.D.

and

W. A. SMITH

British Museum (Natural History)

FRESHWATER BIOLOGICAL ASSOCIATION
SCIENTIFIC PUBLICATION No. 18
Second edition 1974

FOREWORD

Since the copepods have long been a group whose identification is difficult to all but the specialist, this key will be particularly welcome to those whose researches in fresh waters bring them in contact with these almost ubiquitous animals. It is perhaps significant that the authors of this key have not found it possible to construct a key of the usual dichotomous pattern and have therefore resorted to one of the 'tabular' type. In other respects the arrangement follows the general lines of the other keys in the Association's series.

THE FERRY HOUSE,　　　　　　　　　　　　　　H. C. GILSON,
October 1959.　　　　　　　　　　　　　　　　　　　*Director.*

FOREWORD TO SECOND EDITION

This edition is substantially the same as the first edition but the authors have taken the opportunity to make some small improvements to Table 2 and to add an index.

THE FERRY HOUSE,　　　　　　　　　　　　　　E. D. LE CREN
November 1973.　　　　　　　　　　　　　　　　　*Director.*

SBN 900386 20 7

INTRODUCTION

The purpose of this key is to enable the non-specialist to identify British species of free-living fresh- and brackish-water copepods. It does not in any way detract from the value of Gurney's comprehensive Monograph published by the Ray Society in 1933. Gurney's classification and nomenclature have been closely adhered to. Most of the data used in the key are derived from Gurney's work, but they have also been tested on all the species available in the British Museum (Natural History) and amplified accordingly. Thanks are due to the Ray Society for permission to reproduce Gurney's figures, and with few exceptions these are used to illustrate the key.

It is at first a little surprising that Gurney's own keys to the identification of the species have proved difficult to use in practice. This is partly because of the abundance of his drawings, so that any particular detail is difficult to find, but we have also found that a dichotomous key is not practicable for cyclopids and diaptomids because many of them are very close to one another taxonomically and at the same time rather variable, so that a single character difference is often inadequate for their separation. We have, instead, used a tabular form in which taxonomically useful characters are compared in all the species. The variants of each character are indicated on the tables by small letters in accordance with the explanations beginning on pages 15 and 25. Where a character is very variable in a particular species two or more letters may be used. Two letters are also used when a species has a character which is intermediate between the two forms specified. The most reliable characters are given in bold type.

The key is for the identification of calanoids of both sexes and free-living adult female cyclopids. The beginner is advised to start with cyclopids, confining his attention to copepods carrying two egg-sacs, as female calanoids and harpacticoids carry only one, and the presence of egg-sacs ensures that the female is fully adult. Immature specimens have fewer segments to the antennules and sometimes to the legs. The males differ from the females in having prehensile antennules with fewer segments, in having the rudiments of a sixth pair of legs and sometimes in having rather different furcal rami; they are also rather smaller. In spite of these differences it is usually possible to identify the males by comparing them with adult females from the same sample.

Many of the characters may be seen in the intact animal, either alive in a compressorium or as preserved material cleared in lactic acid, but it is necessary to dissect off the legs if the setae and spines are to be seen properly. Lactic acid is a convenient liquid to dissect in, and polyvinyl lactophenol deeply coloured with lignin pink or chlorazol black is a very convenient mounting medium which stains the chitin at the same time as it clears the tissues. The copepod should be dissected under a binocular microscope either by hand or with the aid of a Labgear-Harding micro-dissector. In either case fine tungsten needles are very useful. These are made by dipping the end of a short piece of tungsten wire about 0·2 mm thick (No. 35 or 36 S.W.G.) into a little fused sodium nitrite ($NaNO_2$). The nitrite should be heated in a very small iron or brass container over a small flame; a spirit lamp is convenient. Care should be taken not to spill the molten salt on one's hands. If the nitrite is hot enough the tip of the tungsten wire becomes incandescent and erodes away rapidly leaving a very fine point.

The fifth legs are important, but in cyclopids they are very small and should not be dissected off but left attached to the hinder part of the body, which may be separated off at the movable joint immediately behind the fourth pair of legs. This part of the body should be mounted with the ventral side towards the cover glass; this will make it easy to see the fifth pair of legs and also the furcal rami. The fourth, third, second and first pairs of legs are then taken off in turn from the thorax, care being taken to mount them in order. Apart from the antennules no further dissecting is required, but it is advisable to mount the cephalothorax with the remaining appendages on the slide at the same time. With care all the parts may be mounted in order under the same cover slip. This is most easily achieved by arranging the parts as they are dissected off in small drops of polyvinyl lactophenol on the cover glass and turning this over on to more medium on the slide after the drops have dried.

INTRODUCTION

The beginner is recommended, having mounted a specimen in this manner, to proceed as suggested at the head of Table 1 (pp. 22–3) or Table 2 (pp. 38–41). As experience is gained it will probably be found unnecessary, since many characters are the same for several species, to prepare a complete formula for every specimen. Thus in the case of many cyclopoids it will be quickest to prepare a formula of the 'strong' characters (those in bold type in Table 2) first. This will often serve to show that the specimen belongs to one of a small group of species. Only the 'weaker' characters serving to separate these species need then be studied.

MAJOR GROUPS

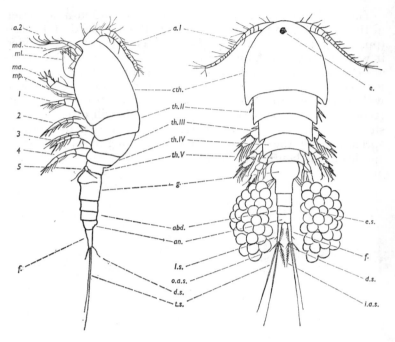

Fig. 1. A female cyclopid copepod in side and dorsal view.

a.1,	antennule	*f.*,	furcal ramus	*o.a.s.*,	outer apical spine
a.2,	antenna	*g.*,	genital segment	*th. II-V.*,	second to fifth thoracic segments
abd.,	abdomen	*i.a.s.*,	inner apical seta		
an.,	anal segment	*l.s.*,	lateral seta	1–5,	pairs of legs carried by the corresponding thoracic segments
cth.,	cephalothorax*	*ma.*,	maxilla		
d.s.,	dorsal seta	*md.*,	mandible		
e.,	eye	*ml.*,	maxillule		
e.s.,	egg sac	*mp.*,	maxillipede	*t.s.*,	terminal seta

*The cephalothorax is formed of the head fused with the first thoracic segment; it carries the antennae, the mouth parts and the first pair of legs.

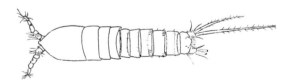

Fig. 2. *Canthocamptus pygmaeus*: male in dorsal view.

I. DICHOTOMOUS KEY TO THE MAJOR GROUPINGS AND CERTAIN GENERA

1   Cephalothorax markedly broader than the abdomen (figs. 1, 6a, b), and distinctly separated from it—   **2**

—   Cephalothorax about the same width as the abdomen, and not distinctly separated (fig. 2). Antennules very short, never exceeding 10 segments. Length of body not exceeding 1 mm—   HARPACTICOIDA
(not included in this key)

2   Antennules with 17–25 segments. Antennae biramous. Egg sacs (if any) single—   CALANOIDA **3**

—   Antennules with 6–17 usually well defined segments. Antennae (fig. 1, a.2) uniramous (exopod absent). Egg sacs paired—   CYCLOPOIDA (p. 25)

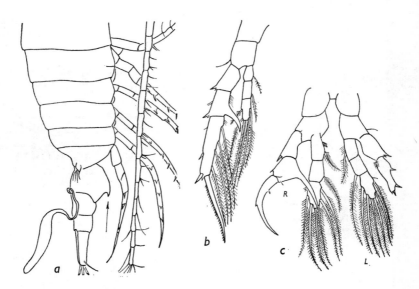

Fig. 3. *Centropages hamatus:*   *a,* female in side view;   *b,* female leg V; *c,* male leg V.

R, right;   *L,* left.

CALANOIDA

3   Antennules with 17–18 segments, sometimes ill defined. Antennae biramous. Female leg V as in fig. 4g; male as in figs. 4d, f. Brackish water or estuarine—
genus ACARTIA 7

—   Antennules with 22–25 segments— 4

4   Furcal ramus short, less than 4 times as long as broad— 5

—   Furcal ramus long and slender, 4-8 times as long as broad— 6

5   Female genital segment with ventral backwardly curved hook (fig. 3a′). Female leg V as in fig. 3b; male fig. 3c—
Centropages hamatus (Lillj.)

—   Female genital segment without a hook. Female leg V as in fig. 7; male as in fig. 9—   genus DIAPTOMUS (p. 15)

6 (4)* Leg V uniramous (female fig. 5a; male figs. 5b, d)—
genus EURYTEMORA 9

—   Leg V biramous (female fig. 4a; male fig. 4b)—
Limnocalanus macrurus Sars

---

* Where a couplet in the key is not reached from the preceding couplet the number of the couplet from which the direction came is indicated thus in parentheses.

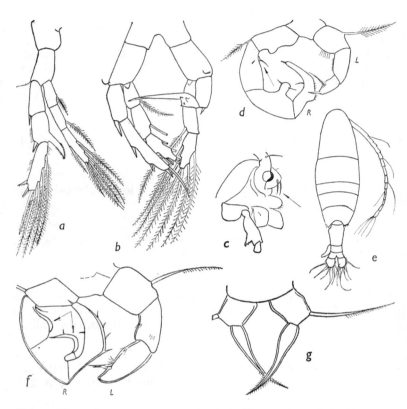

Fig. 4. *Limnocalanus macrurus:* a, female leg V; b, male leg V.
Acartia bifilosa: c, female head in side view.
A. discaudata: d, male leg V; e, female in dorsal view showing spermatophore.
A. clausi: f, male leg V; g, female leg V.

R, right; L, left.

Genus ACARTIA

7 (3) Rostral filaments present (fig. 4c′)—
  Acartia bifilosa Giesbr.

— Rostral filaments absent— 8

8 Female furcal ramus longer than broad; setae not swollen at the base. Male right leg V, segments 2 and 3 each with a prominent inner lobe (fig. 4f′′); segment 4 with 4 spines on the outer margin and one inner spine (fig. 4f)—
  A. clausi Giesbr.

— Female furcal ramus as broad as long; setae swollen at the base (fig. 4e). Male right leg V, segment 2 without a prominent inner lobe but merely a small knob bearing a seta (fig. 4d′); segment 4 with denticles on the outer margin (fig. 4d)— **A. discaudata** Giesbr.

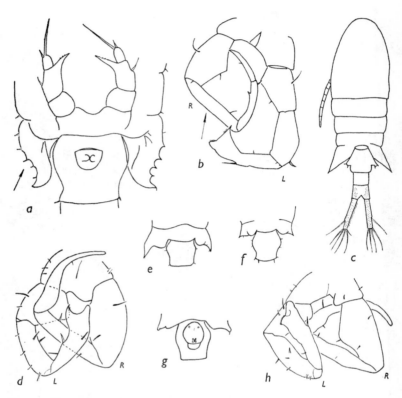

Fig. 5. *Eurytemora velox:* a, female thoracic somite V and abdominal somite I in ventral view, showing leg V and the form of the wings and genital operculum; b, male leg V.
E. affinis: c, female in dorsal view; d, male leg V.
E. americana: e-g, female thoracic somite V and genital segment, e-f, dorsal, g, ventral view; h, male leg V.

R, right; L, left.

Genus EURYTEMORA*

9 (6) Female thoracic somite V with lateral expansions, the outline S-shaped when viewed dorsally, and the outer margin bearing about 5 small hairs (fig. 5a′). Furcal ramus about 4 times as long as broad, the dorsal surface hairy. Male right leg V with a ramus of 4 segments (fig. 5b′). Furcal ramus 6 or more times as long as broad—
Eurytemora velox (Lillj.)

— Female thoracic somite V produced into wings, not S-shaped in outline. Furcal ramus 5–8 times as long as broad. Male right leg V with a ramus of 3 segments (fig. 5d)—   **10**

10 Female thoracic somite V produced into long pointed wings ending in a small spine (fig. 5c). Ramus of male right leg V with the last segment dilated at the base (fig. 5d). Furcal ramus with some dorsal spinules—
E. affinis (Poppe)

— Female thoracic somite V produced into wings of variable shape, but always with a large rounded inner lobe (figs. 5e–g). Ramus of male right leg V with the last segment scarcely dilated at the base (fig. 5h). Furcal ramus without dorsal spinules—   **E. americana** Williams

---

* Gurney (1933, p. 212) states: 'Brady has also recorded (1910) *E. lacustris* from Lea Mills Pond near Sheffield. Prof. Cannon has kindly obtained samples from this pond, which now contains no *Eurytemora*'. As Brady's record of this species is merely given in a list, and as Gurney dismisses it without further comment and we have not been able to find any subsequent record of its occurrence in Britain, the species has been omitted.

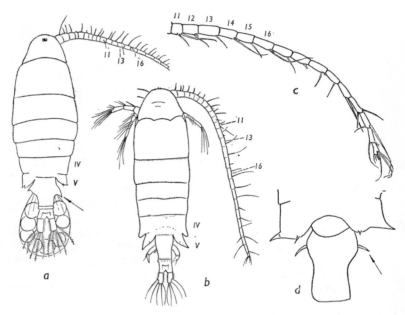

Fig. 6. *Diaptomus*: *a*, *D. castor* female bearing resting eggs in dorsal view; *b*, *D. laciniatus* female in dorsal view; *c*, *D. wierzejskii* female, antennule segments 11–25; *d*, *D. gracilis* female thoracic somite V in dorsal view.
IV, V, thoracic somites;   11–16, segments of the antennule.

II. FORMULA KEY TO *DIAPTOMUS*

Characters used in Table I

Part 1. Females

A Antennule: numbers of setae on segments 11, 13 and 16.

	Segment 11	Segment 13	Segment 16	
a	2	2	2	(fig. 6a)
b	2	1	2	(fig. 6b)
c	2	1	1	(fig. 6c)
d	2	2	1	
e	1	1	1	

B Genital somite: shape
   a A large triangular process on each side (fig. 6a↗).
   b A spine on each side (fig. 6d↗).
   c No processes or spines (fig. 6b).

C Thoracic somites IV and V:
   a A wing-like lateral process on both somites (fig. 6b).
   b No such process on somite IV (fig. 6a).

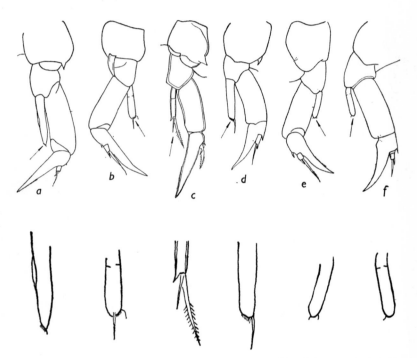

Fig. 7. *Diaptomus*, female leg V, detail of endopods below; *a*, *D. laciniatus*; *b*, *D. gracilis*; *c*, *D. castor*; *d*, *D. vulgaris*; *e*, *D. wierzejskii*; *f*, *D. laticeps*.

D Leg V: number of segments in the endopod
   a one, or if two they are imperfectly defined (figs. 7a, b).
   b two distinct segments (fig. 7c).

E Leg V: length of the endopod compared with that of the first segment of the exopod
   a the same or nearly so (figs. 7a, d).
   b more than half the length but not more than three-quarters (fig. 7b).
   c half the length or less (fig. 7e).

F Leg V: armature of the endopod
   a One or two minute apical spines or setae without a fringe of hairs (figs. 7e′, f′).
   b A small point and a fine fringe of hairs (fig. 7a′).
   c One or two small spines or setae and a terminal fringe of hairs (figs. 7b′, d′).
   d Two unequal setae and a small spine on the inner angle (fig. 7c′).

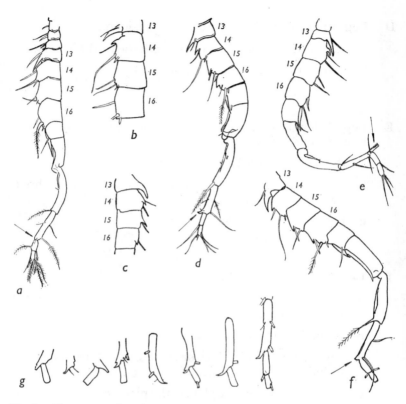

Fig. 8. *Diaptomus*, male prehensile antennule (segments numbered): *a*, *D. castor*; *b*, *D. laticeps*, segments 13-16; *c, d*, *D. vulgaris*; *e*, *D. wierzejskii*; *f*, *D. gracilis*; *g*, *D. laticeps*, end of antennule showing the terminal process on the antepenultimate segment.

Part 2. Males

G **Prehensile antennule:** spinous or tooth-like processes on segments 14–16 (Segment 13 always bears a conspicuous spine).

	Segment 14	Segment 15	Segment 16	
a	0	0	0	(fig. 8a)
b	1	0	0	(figs. 8b, e)
c	1	1	0	(fig. 8c)
d	1	1	1	(figs. 8d, f)

H **Prehensile antennule:** antepenultimate segment
  a without any distal process (fig. 8a♂).
  b with a short hook (figs. 8d♂, f♂).
  c typically shaped as a bill-hook about $\frac{1}{3}$ of the length of segment 24, but sometimes straight, or bifid at the tip, or with a small tooth on the margin (fig. 8g).
  d hook-shaped with a conspicuous saw edge (fig. 8e♂).

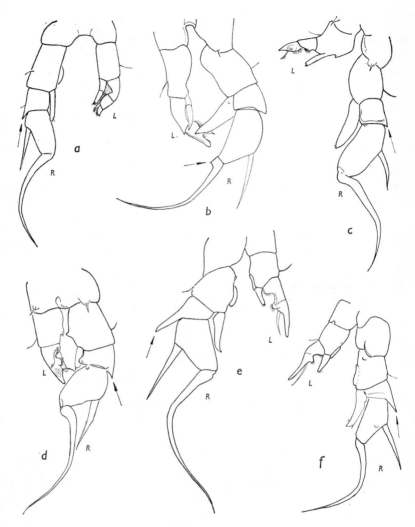

Fig. 9. *Diaptomus*, male leg V: *a*, *D. castor*;   *b*, *D. laciniatus*; *c*, *D. gracilis*;   *d*, *D. vulgaris*;   *e*, *D. laticeps*;   *f*, *D. wierzejskii*. *R*, right;   *L*, left.

J Left leg V: terminal process of exopod
   a short, pointed, curving outwards, inner margin hairy; posterior face smooth (fig. 9a).
   b short, blunt, with serrated edge; posterior face smooth (fig. 9b).
   c short, blunt, with a curved inner seta; posterior face hairy (figs. 9c, d).
   d long, finger-like, with a long slender inner spine; posterior face smooth (figs. 9e, f).

K Left leg V: endopod
   a with two segments (figs. 9a, b).
   b with one segment (figs. 9c, e, f).

L Right leg V: outer angle of exopod segment 1
   a produced into a small blunt projection, or only slightly produced (figs. 9b, c′).
   b produced into a sharp point (figs. 9a′, d′).
   c produced into a strong spinous projection (figs. 9e′, f′).

M Right leg V: terminal claw of exopod 2
   a A conspicuous square, pointed projection at the base (fig. 9b′).
   b Base without any projection.

N Right leg V: length of endopod compared with exopod 1
   a Endopod distinctly longer than both inner and outer margins of exopod 1 (figs. 9b, c).
   b Endopod much exceeding the inner but about equalling the outer margin of exopod 1 (figs. 9e, f).
   c Endopod about equal to both margins (fig. 9d).

22   TABLE I — *DIAPTOMUS*

Table 1. *DIAPTOMUS*

For the meanings of the symbols used to indicate the characters of the species, see pp. 15-21. The most reliable characters are shown in bold type.

It is helpful to write the letters A to N on a sheet of paper using the same spacing as printed below. Fill in the most easily observed characters first and holding the sheet against the table eliminate some species and decide which will be useful characters for further sorting. Continue in this way until only one possibility remains.

		D. castor	*D. laciniatus*	*D. gracilis*	*D. vulgaris*	*D. laticeps*	*D. wierzejskii*
Females							
A	Antennule, setae on segments 11, 13 and 16	a	b	e	e	d	c
B	Shape of genital somite						
C	Thoracic somites IV and V	**a**	c	b	b	b	b
D	Leg V: number of segments in endopod	**a**	**a**	b	b	b	b
E	Leg V: endopod compared with first exopod segment	b	a	a	a	ab	a
F	Leg V: armature of endopod	d	b	c	c	a	a

TABLE I — *DIAPTOMUS*

(Table 1 continued)

Males		*D. castor*	*D. laciniatus*	*D. gracilis*	*D. vulgaris*	*D. laticeps*	*D. wierzejskii*
G	Prehensile antennule: spinous processes on segments 14-16	a	b	d	cd	b	b
H	Prehensile antennule: antepenultimate segment	a	a	d	ab	c	**d**
J	Left leg V: terminal process of exopod	a	b	c	c	d	d
K	Left leg V: endopod	a	ab	ab	b	b	b
L	Right leg V: outer angle of exopod segment 1	b	a	ab	b	c	c
M	Right leg V: terminal claw of exopod 2	b	**a**	b	b	b	b
N	Right leg V: length of endopod compared with exopod 1	b	a	a	c	b	b

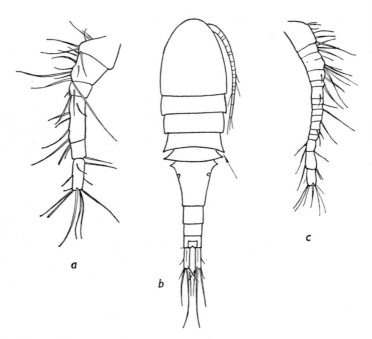

Fig. 10. *a*, *Halicyclops christianensis*, antennule; *b*, *Cyclops scutifer*, dorsal view; *c*, *C. languidus*, antennule.

III. FORMULA KEY TO MATURE FEMALE CYCLOPOIDA

Characters used in Table 2

A **Body:** length measured from the anterior margin of the cephalothorax to the posterior end of the furcal ramus, excluding the furcal setae (fig. 1)
    a not exceeding 0·9 mm.
    b 0·9 – 1·5 mm.
    c 1·5 mm or more.

B **Antennule:** number of segments
    a 6 (fig. 10a).   e 12.
    b 8.   f 14.
    c 10.   ġ 16 (fig. 10c).
    d 11.   h 17.

C **Antennule:** length
    a not reaching to the end of the cephalothorax.
    b about the same as the cephalothorax (fig. 1).
    c reaching beyond the cephalothorax (fig. 10b).

D **Antennule:** terminal five segments
    a The last two segments together shorter than the previous three together (fig. 10a).
    b The last two segments together longer than the previous three (fig. 10c).

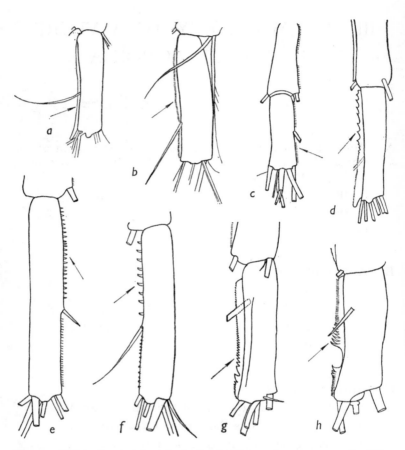

Fig. 11. *Cyclops*, female antennules to show hyaline membranes: *a*, *C. albidus*, segment 17; *b*, *C. prasinus*, segment 12; *c*, *C. strenuus abyssorum*, segments 16 and 17; *d*, *C. fuscus*, segment 17; *e*, *C. macruroides* (s. str.), segment 12; *f*, *C. macruroides denticulatus*, segment 12; *g, h*, *C. leuckarti*, segment 17.

E **Antennule:** hyaline membrane on the last segment

   a absent.

   b smooth, or with very fine hairs (figs. 11$a\prime$, $b\prime$, $c\prime$).

   c with a serrated edge (figs. 11$d\prime$, $g\prime$, $h\prime$).

   d broken up in its proximal half into 15 or more separate spinules (fig. 11$e\prime$).

   e represented by a row of less than 12 spinules in its proximal half (fig. 11f).

Since it is very delicate and transparent, this membrane may be very difficult for the inexperienced to see. An oil-immersion lens and careful illumination and focussing are often necessary. The membrane may stand out as figured, but it may also come to lie under the antennule or between it and the coverslip.

F Legs I-IV: number of spines on the last segment of the exopod

	Legs				
	I	II	III	IV	
a	2	3	3	3	
b	3	3	3	3	(fig. 12)
c	3	4	3	3	
d	3	4	4	3	
e	3	4	4	4	
f	4	4	4	3	

Fig. 12. *Cyclops languidus*: *a*, leg I; *b*, leg II; *c*, leg III; *d*, leg IV. *end.*, endopod, the inner branch of the leg; *exp.*, exopod, the outer branch. Capital letters refer to the paragraphs on pp. 28–31.

G Legs I-IV: number of segments in the outer (exopod) and inner (endopod) branches

	Legs								
	I		II		III		IV		
	ex.	end.	ex.	end.	ex.	end.	ex.	end.	
a	2	2	2	2	2	2	2	2	
b	2	2	2	2	2	2	3	2	
c	2	2	3	2	3	3	3	3	(fig. 12)
d	3	3	3	3	3	3	3	3	

CYCLOPOIDA

H Legs I-IV: inner seta of the first segment of the exopod
   a present on all four legs (fig. 12).
   b absent on all four legs.
   c present on legs I–III; absent on leg IV (fig. 13*a*).

K Legs I-IV: number of setae on the last segment of the exopod

	Legs				
	I	II	III	IV	
a	4	4	4	3	
b	4	4	4	4	
c	4	4	4	5	
d	5	4	4	4	(fig. 12)
e	5	5	5	4	
f	5	5	5	5	

L Leg IV, endopod: apical spines of the last segment
   a One apical spine (figs. 13*a, c*).
   b Two apical spines, the outer less than half the length of the inner (fig. 13*b*).
   c Two apical spines, the outer more than half but not exceeding the length of the inner (fig. 12*d*).
   d Two apical spines, the outer longer than the inner (fig. 13*d*).

  The proportions of these characters are very variable in some species. Measurements must be made with care using camera lucida and dividers.

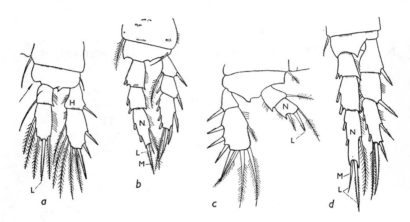

Fig. 13. *Cyclops*, leg IV: a, *C. minutus*; b, *C. vicinus*; c, *C. unisetiger*; d, *C. fuscus*.

Capital letters refer to the paragraphs on pp. 29 and 31.

M **Leg IV, endopod:** length of the inner apical spine
   a longer than the last segment.
   b about the same length as the segment.
   c shorter than the segement (figs. 12d, 13b, d).

N **Leg IV, endopod:** proportions (greatest length and breadth) of the last segment
   a Less than $1\frac{1}{2}$ times as long as wide (fig. 13c).
   b between $1\frac{1}{2}$ and 2 times as long as wide.
   c Between 2 and $2\frac{1}{2}$ times as long as wide.
   d Between $2\frac{1}{2}$ and 3 times as long as wide (fig. 13b).
   e More than 3 times as long as wide (fig. 13d).

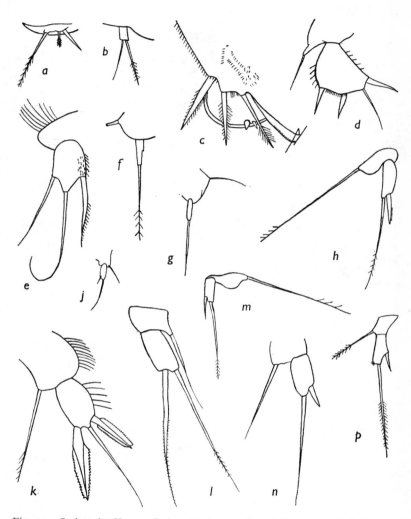

Fig. 14. *Cyclops*, leg V: a, *C. demetiensis*; b, *C. unisetiger*; c, *C. phaleratus*; d, *Halicyclops christianensis*; e, *Cyclops prasinus*; f, *C. varicans* (s. str.); g, *C. varicans rubellus*; h, j, *C. nanus*; k, *Cyclopina norvegica*; l, *Cyclops hyalinus*; m, *C. sensitivus*; n, *C. vernalis*; p, *C. strenuus abyssorum*.

CYCLOPOIDA

P **Leg V:** number of segments
  a one, hardly demarcated from the somite (figs. 14*a*, *c*).
  b one, distinctly separated from the somite (figs. 14*b*, *e*).
  c two, the first fused with the somite (indicated by its seta) (figs. 14*d*, *f*, *g*, *j*).
  d two, distinctly separated from the somite (figs. 14*h*, *k–p*).

R **Leg V:** armature
  a 4 or 5 setae or spines on the last segment (fig. 14*d*).
  b 3 setae or spines on the last segment (figs. 14*a*, *c*, *e*, *k*).
  c 2 long setae or spines on the last segment (figs. 14*b*, *l*).
  d 1 apical seta with an inner spinule as long as or longer than the segment (fig. 14*m*).
  e 1 apical seta with an inner spinule shorter than the segment and inserted near its apex (figs. 14*h*, *j*, *n*).
  f 1 apical seta with an inner spinule shorter than the segment and inserted near its middle (figs. 14*f*, *p*).
  g 1 apical seta with no spinule (fig. 14*g*).

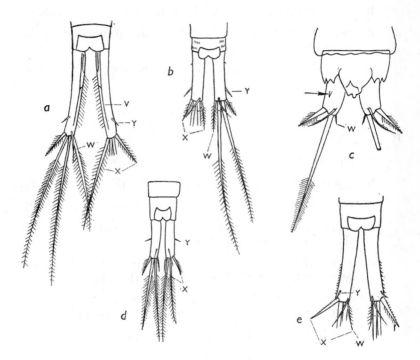

Fig. 15. *Cyclops*, furcal rami in dorsal view: *a*, *C. vicinus*; *b*, *C. vernalis*; *c*, *C. unisetiger*; *d*, *C. minutus*; *e*, *C. agilis*.

Capital letters refer to the paragraphs on p. 35.

CYCLOPOIDA

S **Furcal ramus:** approximate ratio of length: breadth
    a 8 : 1 (fig. 15a)   **e** 3 : 1
    b 6 : 1.   **f** $2\frac{1}{2}$: 1.
    c 5 : 1.   **g** 2 : 1 (fig. 15c).
    d 4 : 1.   **h** $1\frac{1}{2}$ or less : 1.

T **Furcal ramus:** inner margin
    a hairy (fig. 15a).   **b** smooth (figs. 15b–e).

U **Furcal ramus:** outer margin
    a with longitudinally arranged denticles or spinules (fig. 15e).
    b smooth, or with denticles not arranged in a longitudinal row (figs. 15a–d).

V **Furcal ramus:** dorsal ridge
    a present (fig. 15a).   **b** absent (figs. 15b–e).

W **Furcal ramus:** inner apical seta
    a minute or absent (fig. 15c).
    b shorter than the ramus (figs. 15b, d, e).
    c longer than the ramus (fig. 15a).

X **Furcal ramus:** relative lengths of the apical setae
    a Inner shorter than the outer (fig. 15d).
    b About the same length (fig. 15e).
    c Inner longer, but not twice as long as the outer (fig. 15b).
    d Inner twice as long.
    e Inner more than twice as long (fig. 15a).

Y **Furcal ramus:** position of lateral seta.
    a In the middle of the ramus (fig. 15d).
    b Near the end or in the distal third of the ramus (figs. 15a, b, e).

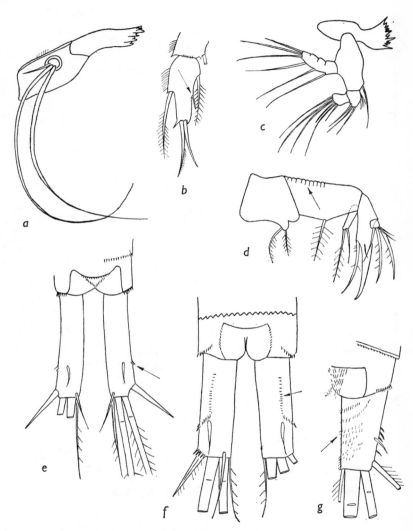

Fig. 16. *a*, *Cyclops albidus*, mandible; *b*, *C. albidus*, leg IV, distal segment of endopod; *c*, *Cyclopina norvegica*, mandible; *d*, *Cyclops leuckarti*, maxilla; *e*, *C. fimbriatus* (s. str.), furcal rami; *f*, *C. fimbriatus poppei*, furcal rami; *g*, *C. phaleratus*, furcal ramus.

CYCLOPOIDA 37

Z Special characters diagnostic for certain species:

Cyclopina norvegica — the mandibular palp is large and two-branched (fig. 16c; contrast with fig. 16a).

Cyclops scutifer and *C. vicinus* — thoracic somites IV and V are expanded laterally and pointed (fig. 10b↗).

Cyclops minutus and *C. leuckarti* — the basis of leg I has no inner seta (fig. 13a↗; contrast fig. 12a↗).

Cyclops albidus — the distal inner seta of the endopod of leg IV is reduced (fig. 16b↗).

Cyclops leuckarti — the outer margin of the basis of the maxilla is conspicuously ribbed (fig. 16d↗).

Cyclops unisetiger — the furcal ramus has only one terminal seta (fig. 15c).

Table 2. CYCLOPOIDA

TABLE 2—CYCLOPOIDA

For the meanings of the symbols used to indicate the characters of the species, see pp. 25-37. The most reliable characters are shown in bold type.

It is helpful to write the letters A to Y on a sheet of paper using the same spacing as printed below. Fill in the most reliable and easily observed characters first, and, holding the sheet against the table, eliminate some species and decide which will be useful characters for further sorting. Continue in this way until only one possibility remains.

	Body	Antennule				Legs I-IV				Leg IV, endopod			Leg V		Furcal ramus						
	Length	No. of segments	Length	Last 5 segments	Hyaline membrane	Spine formula	No. of segments	Exp. 1, inner seta	Setae, last Exp. seg.	Apical spines	Inner ap. spine	Proportions, last seg.	No. of segments	Armature	Length : Breadth	Inner margin	Outer margin	Dorsal ridge	Inner apical seta	Lengths of setae	Lateral seta
	A	B	C	D	E	F	G	H	K	L	M	N	P	R	S	T	U	V	W	X	Y
Genus CYCLOPINA																					
*C. norvegica** ..	a	c	a	a	a	f	d	a	c	c	a	b	**d**	**b**	e	**b**	**b**	**b**	c	c	a
Genus HALICYCLOPS																					
H. christianensis ..	a	**a**	**a**	a	a	**d**	**d**	**a**	f	c	a	bc	**c**	a	g	**b**	**b**	**b**	**a**	a	a
H. neglectus ..	a	**a**	**a**	a	a	**d**	**d**	**a**	f	c	a	a	**c**	a	h	**b**	**b**	**b**	bc	a	a

* See § Z, p. 37, for special characters of this species.

(Table 2 continued)

TABLE 2—CYCLOPOIDA

	A	B	C	D	E	F	G	H	K	L	M	N	P	R	S	T	U	V	W	X	Y
Genus Cyclops																					
Sub-genus Macrocyclops																					
C. fuscus	c	**h**	c	b	**c**	**d**	**d**	**a**	**f**	d	c	e	**d**	**b**	g	**a**	**b**	**b**	b	c	b
C. albidus*	c	**h**	c	b	b	**d**	**d**	**a**	**f**	d	c	d	**d**	**b**	f	**b**	**b**	**b**	b	d	b
C. distinctus	c	**h**	c	b	b	**d**	**d**	**a**	**f**	d	c	d	**d**	**b**	e	**a**	**b**	**b**	b	c	b
Sub-genus Tropocyclops																					
C. prasinus	a	**e**	c	a	b	**d**	**d**	**a**	**f**	b	a	d	**b**	**b**	efg	**b**	**b**	**b**	b	bc	a
Sub-genus Eucyclops																					
C. agilis (s. str.) ..	b	**e**	c	b	b	**d**	**d**	**a**	**f**	c	b	d	**b**	**b**	c	**b**	**a**	**b**	b	bc	b
C. a. speratus	b	**e**	c	b	b	**d**	**d**	**a**	**f**	c	b	d	**b**	**b**	ab	**b**	**a**	**b**	b	c	b
C. macruroides (s. str.)	b	**e**	c	a	**d**	**d**	**d**	**a**	**f**	c	b	d	**b**	**b**	ab	**b**	**a**	**b**	b	c	b
C. m. denticulatus ..	b	**e**	c	a	**e**	**d**	**d**	**a**	**f**	c	c	d	**b**	**b**	abc	**b**	**a**	**b**	b	cd	b
C. macrurus	b	**e**	c	a	b	**d**	**d**	**a**	**f**	c	c	d	**b**	**b**	a	**b**	**b**	**b**	b	d	b
Sub-genus Paracyclops																					
C. fimbriatus (s. str.) ..	a	**b**	a	a	a	**d**	**d**	**a**	**f**	c	a	b	**b**	**b**	bcd	b	b[1]	**b**	b	c	b
C. f. poppei	a	**b**	a	a	a	**d**	**d**	**a**	**f**	c	a	b	**b**	**b**	e	b	b[2]	**b**	b	c	b
C. affinis	a	**d**	a	a	a	bc	**d**	**a**	**f**	b	a	b	**b**	**b**	fg	**b**	**b**	**b**	b	a	b
Sub-genus Ectocyclops																					
C. phaleratus	b	c	a	a	a	**d**	**d**	**a**	**f**	b	a	ab	a	**b**	g	a[3]	b	**b**	b	abc	b

* See § Z, p. 37, for special characters of this species.

[1] A few transversely arranged spinules near the lateral seta (Fig. 16e1).
[2] With a longitudinal row of spinules dorsally, but not laterally (Fig. 16f1).
[3] These hairs are the continuations of transverse rows (Fig. 16g1).

40 TABLE 2—CYCLOPOIDA

(Table 2 continued)

	A	B	C	D	E	F	G	H	K	L	M	N	P	R	S	T	U	V	W	X	Y
Sub-genus CYCLOPS																					
C. strenuus (s. str.)	c	h	c	b	b	c	d	a	f	c	c	d	d	df	b	a	b	a	b	c	b
C. s. abyssorum	b	h	c	b	b	c	d	a	f	b	a	de	d	f	b	a	b	a	c	c	b
C. scutifer*	b	h	c	b	b	c	d	a	f	b	a	d	d	f	d	a	b	a	e	e	ab
C. furcifer	c	h	c	b	b	ac	d	a	f	b	a	d	d	f	a	a	b	a	b	c	b
C. vicinus*	c	h	c	b	b	a	d	a	f	b	c	d	d	f	a	a	b	a	c	e	b
Sub-genus ACANTHOCYCLOPS																					
C. viridis	c	h	b	b	a	a	d	a	b	c	c	c	d	f	bcd	a	b	b	c	de	b
C. gigas (s. str.)	c	h	b	b	a	a	d	a	b	c	bc	d	d	f	b	a	b	b	bc	c	b
C. g. latipes	c	h	a	a	a	a	d	a	b	c	c	b	d	f	bc	a	b	b	c	c	b
C. vernalis (s. str.)	bc	h	a	b	a	ae	d	a	b	d	c	c	d	e	bc	b	b	b	bc	bc	b
C. v. americanus	b	h	a	b	a	ae	d	a	b	c	c	d	d	e	b	b	b	b	b	d	b
C. venustus	b	e	a	a	a	e	e	c	f	c	c	b	d	f	de	a	b	b	bc	d	b
C. sensitivus	a	h	a	b	a	a	d	b	b	c	c	bc	d	d	f	a	b	b	e	e	a
C. bicuspidatus (s. str.)	b	h	b	b	a	a	d	a	b	cd	c	cd	d	d	b	b	b	b	b	bc	ab
C. b. lubbocki	b	f	b	b	a	a	d	a	b	cd	c	cd	d	d	b	b	b	b	b	bc	a
C. b. thomasi	b	h	c	b	a	a	d	a	b	d	c	e	d	d	b	b	b	b	bc	bc	a
C. bisetosus	b	h	b	b	a	a	d	a	b	c	c	a	d	d	bc	b	b	b	b	a	b
C. crassicaudis	b	e	b	a	a	a	d	a	b	c	b	a	d	d	cd	b	b	b	b	a	b
C. languidus	ab	g†	b	b	a	b	c	a	d	c	c	b	d	d	cd	b	b	b	b	a	b

* See § Z, p. 37, for special characters of these species.
† Segment 3 may be partially divided (Fig. 10c†).

(Table 2 continued)

TABLE 2—CYCLOPOIDA

Sub-genus	A	B	C	D	E	F	G	H	K	L	M	N	P	R	S	T	U	V	W	X	Y
ACANTHOCYCLOPS continued																					
C. languidoides (s. str.)	a	**d**	b	a	a	**b**	c	**a**	**d**	c	c	a	**d**	**d**	c	**b**	**b**	**b**	b	a	b
C. l. hiberniae	ab	**d**	ab	a	a	**b**	c	**a**	**d**	c	c	a	**d**	**d**	b	**b**	**b**	**b**	b	a	b
C. l. hypnicola	a	**d**	ab	a	a	**b**	c	**a**	**d**	c	c	a	cd	**d**	f	**b**	**b**	**b**	b	a	b
C. l. eriophori	a	**d**	ab	a	a	**b**	c	**a**	**d**	c	c	a	cd	**e**	de	**b**	**b**	**b**	b	a	b
C. nanus	a	**d**	a	a	a	**b**	c	**a**	**d**	c	c	c	cd	**e**	c	**b**	**b**	**b**	b	a	a
Sub-genus MICROCYCLOPS																					
C. varicans (s. str.)	a	**e**	**a**	**a**	a	**d**	a	c	**f**	c	bc	cd	c	fg	def	**b**	**b**	**b**	b	c	b
C. v. rubellus	a	de	**a**	**a**	a	**d**	a	c	**f**	c	c	c	c	**g**	ef	**b**	**b**	**b**	bc	b	b
C. bicolor	a	cd	**a**	**a**	a	**d**	a	c	**f**	b	c	c	c	eg	cd	**b**	**b**	**b**	b	e	b
C. minutus*	ab	**d**	**a**	**a**	a	**d**	a	c	**f**	**a**	a	b	c	**d**	cd	**b**	**b**	**b**	**b**	a	a
C. gracilis	a	**d**	**c**	**a**	a	**d**	a	c	**f**	b	a	de	c	e	e	**b**	**b**	**b**	c	a	b
C. unisetiger*	a	**d**	**a**	**a**	a	e	ab	**b**	**d**	**a**	a	a	**b**	c	g	**b**	**b**	**a**	**a**	c	a†
C. demetiensis	a	**d**	**a**	**a**	a	**d**	a	**b**	**a**	**a**	a	a	**a**	**b**	h	**b**	**b**	**b**	b	a	a
Sub-genus MESOCYCLOPS																					
C. leuckarti*	b	**h**	c	b	**c**	**a**	**d**	**a**	**b**	d	c	de	**d**	c	de	b	**b**	**b**	c	e	a
C. hyalinus	ab	**h**	c	b	b	**a**	**d**	**a**	**b**	c	c	cd	**d**	c	fg	b	**b**	**b**	c	e	b
C. dybowskii	a	**h**	c	b	b	**a**	**d**	**a**	**b**	d	c	d	**d**	c	f	b	**b**	**b**	c	c	b

* See § Z, p. 37, for special characters of these species.
† Lateral seta spine-like and inserted on the dorsal side (Fig. 15c1).

DISTRIBUTION AND ECOLOGY

The following notes are based on Gurney's (1933) data for each species, with additions from the collections in the British Museum (Natural History) and from the records of the Freshwater Biological Association compiled by Mr W. J. P. Smyly. Gurney's names are used except for the two species of *Halicyclops*, since, as Fischer's specimens of *H. aequoreus* came from Madeira, his name must apply to them and not to either of the British species. Synonyms are given only when Sars (1913–18) or Kiefer (1929) used a different name; fuller synonymy is given by Gurney (1933).

In giving information on the distributions in the Lake District the following symbols are used:

A	Blea Tarn, Armboth	H	Haweswater
B	Bassenthwaite Lake	Hi	Highlow Tarn, (Tarn Hows)
Bg	Bigland Tarn		
Bl	Blelham Tarn	L	Loughrigg Tarn
Br	Brotherswater	LL	Little Langdale Tarn
BT	Blea Tarn, Langdale	Lp	Loughrigg pools
Bu	Buttermere	Ly	Lily Tarn, Loughrigg
BW	Blea Water, Haweswater	M	Mockerkin Tarn
C	Coniston Water	R	Rydal Water
Cr	Crummock Water		
Cu	Cunswick Tarn	Sm	Small Water, Haweswater
D	Derwentwater	Sy	Sty Head Tarn
Dv	Devoke Water	T	Tewet Tarn, Keswick
E	Ennerdale Water	U	Ullswater
Ea	Easedale Tarn		
Ee	Eel Tarn	W	Windermere
Es	Esthwaite Water	Wa	Wastwater
		WT	Watendlath Tarn, Borrowdale
F	Floutern Tarn		
G	Grasmere	Z	Other waters, including swamps, pools and small tarns.
Go	Goat's Water		
Gr	Greendale Tarn		

CALANOIDA

Centropages hamatus (Lilljeborg). Female, 1·1 – 1·2 mm; male, 1·0 – 1·14 mm. Most abundant between June and September. Males usually exceed females. A marine species, but penetrates into brackish estuarine regions.
Distributed all round the British Isles, but relatively rare in the south. East coast estuaries, tidal rivers in Norfolk, tidal lochs in Scotland, and Shetlands.
Lake District: not recorded.

Limnocalanus macrurus Sars. Female, 2·10 – 2·27 mm; male, 1·85 – 2·16 mm. Found only in deep lakes, and usually only in the deeper layers; a cold water species. The thorax contains large fat globules.
Lake District: Gurney found it to be the most abundant species in the plankton of the deep eastern part of Ennerdale Water, but it could not be found in 1956 and may now be extinct in this country.

Diaptomus castor (Jurine). Female, 1·8 – 2·5 mm; male, 1·75 – 2·3 mm. Shallow ponds or ditches, tolerant of a wide range of conditions, a characteristic species of pools which dry up in summer. Most commonly associated with *Daphnia pulex*. Characteristic of winter months, rare in the summer.
Widely distributed and not uncommon in the east and south of England.
Lake District: Ly.

Diaptomus laciniatus Lilljeborg. Female, 1·05 – 1·2 mm; male, 1·4 – 1·55 mm.
Scotland and Ireland in large bodies of water. N. Wales (Llyn Bodlyn).
Lake District: not recorded.

Diaptomus gracilis Sars. Female, 0·99 – 1·65 mm; male, 0·98 – 1·5 mm. The characteristic *Diaptomus* of large and small lakes frequently found associated with other species. May be found in ponds, but shows distinct preference for open water.
Scotland, Norfolk Broads, Ireland.
Lake District: A Bg BT Bu BW Cu Dv Ea F Gr Hi L LL M Sm St T Wa WT.

Diaptomus vulgaris Schmeil.   Female, 1·5 – 2·0 mm; male, 1·3 – 1·8 mm.
Eastern and southern counties.
Lake District: not recorded.

Diaptomus laticeps Sars.   (*D. bacillifer* Brady, *D. hircus* Brady, Scott).   Female, 1·15 – 1·6 mm; male, 1·05 – 1·4 mm.
Scotland, Ireland, Wales, N. England.
Lake District: Go H W and nowhere else.

Diaptomus wierzejskii Richard.   (*D. serricornis* Brady). Female, 1·54 – 1·8 mm; male 1·2 – 1·45 mm.   Confined to oligotrophic lakes in the north.
Scotland and Ireland; Devon.
Lake District: not recorded.

Eurytemora velox (Lilljeborg).   (*E. clausi* Brady).   Female, 1·6 – 1·95 mm; male, 1·3 – 1·65 mm.   Tolerates a wide range of salinity.
Scotland, Ireland, Wales, S. and E. England.
Lake District: not recorded.

Eurytemora affinis (Poppe).   (*E. hirundoides* (Nordquist), Sars, auct.).   Female, 1·2 – 1·73 mm; male, 1·36 – 1·65 mm. Widely distributed in brackish waters.
England and S. Scotland.
Lake District: not recorded.

Eurytemora americana Williams.   (*E. Thompsoni* Lowndes). Female, 1·11 – 1·82 mm; male, 1·04 – 1·53 mm.   Brackish pools.
Sussex; Isle of Wight.
Lake District: not recorded.

Acartia clausi Giesbrecht (*Dias longiremis* Brady, *Acartia ensifera* Brady).   Female, brackish water form, 0·70 – 0·82 mm; North Sea form, 1·47 mm; male, brackish water form, 0·71 mm; North Sea form, 1·31 mm.
Small form N. Uist and Norfolk.   Large form common all round British coasts.
Lake District: not recorded.

Acartia discaudata Giesbrecht. Female, 1·0 – 1·2 mm; male, 0·9 – 1·1 mm. River mouths and harbours.
England, Ireland, Scotland.
Lake District: not recorded.

Acartia bifilosa Giesbrecht. Female, 0·77 – 0·9 mm; male, 0·77 – 0·86 mm. Brackish water and estuarine.
N. Uist, Norfolk, Tamar estuary.
Lake District: not recorded.

CYCLOPOIDA

Cyclopina norvegica Boeck. Female 0·58 – 0·62 mm. A marine littoral species common in the estuaries of the east coast.
England, Scotland and Ireland.
Lake District: not recorded.

Halicyclops christianensis Boeck (*H. aequoreus* auct. not Fischer; *H. magniceps* Sars). Female, 0·65 – 0·73 mm. On each side posteriorly of the genital somite is a faintly marked 'lateral disc'. Egg sacs closely pressed to abdomen. Colour greyish white, sometimes with rosy red. Brackish coastal pools with little vegetation.
Generally distributed in the British Isles.
Lake District: not recorded.

Halicyclops neglectus Kiefer (*H. aequoreus propinquus*). Female, 0·53 – 0·72 mm. Distribution as for *H. christianensis*.

Sub-genus MACROCYCLOPS

Cyclops fuscus (Jurine) (*Pachycyclops signatus*). Female, 2·0 – 2·5 mm. Occasionally nearly colourless, but generally richly coloured, and appearing almost black to the naked eye; ramus and abdomen bluish-green; thorax dark green, with blue patches; receptaculum brick-red. Found only in clear, weedy waters. Present throughout the year, but much commoner in summer.

Generally distributed throughout the British Isles, rarer in the north (Scotland) than in the south and east. Not uncommon in Ireland.

Lake District: B(?) Bg Bl BT C Cu Ea Es G Gr Hi L LL Lp Ly M R St Z.

Cyclops albidus (Jurine) (*Pachycyclops annulicornis*). Female, 1·5 – 2·5 mm. Egg sacs large, divergent. Generally greyish, with darker bands across the thorax. Found throughout the year, but very much more abundant in summer.

Generally distributed throughout British Isles. Very common in Scottish lochs.

Lake District: B Bg Bl Br BT BW C Cr Cu D Dv E Ea Es G L LL Ly R Sm St U W Wa Z.

Cyclops distinctus (Richard) (*Pachycyclops bistriatus*). Female, 2·0 mm. Blue, with greenish markings, the end of the abdomen and antennules purple. Egg sacs large, not very divergent. Often found with *C. fuscus* and *C. albidus*.

Generally distributed in England and Scotland.

Lake District: not recorded.

Sub-genus TROPOCYCLOPS

Cyclops prasinus (Fischer, Schmeil). Female, 0·68 – 0·75 mm. Eye very large. Egg sacs closely apposed to abdomen. Colour dark green. A swimming rather than a creeping form; swims on its back. A summer form, breeding in June and September, found most commonly in rather small ponds. Not planktonic in this country.

Generally distributed, commonest in South of England, rare in Scotland.

Lake District: Ee Gr Ly WT Z.

Sub-genus EUCYCLOPS (LEPTOCYCLOPS)

Cyclops agilis (s. str.) (Koch, Sars) (*Eucyclops serrulatus* auct; *Leptocyclops agilis*). Female, 0·8 – 1·45 mm. Cuticle often

conspicuously pitted. Egg sacs rather divergent, compact and pointed at end. Colour variable, from colourless to deep yellow.
Generally distributed in the British Isles.
Lake District: A B Bg Bl Br BT Bu BW C Cr Cu D Dv E Ea Es F G Go Gr H Hi L LL Lp Ly M R Sm St T U W Wa WT Z.

Cyclops agilis speratus (Lilljeborg). Female, 1·0 – 1·42 mm.
Generally distributed in the British Isles.
Lake District: B Bg Bu C Ea Es Go L LL R W Z.

Cyclops macruroides (s. str.) (Lilljeborg). Female, 1·04 – 1·25 mm.
England, Scotland.
Lake District: A BL Br Cu Dv Ea Es Hi St W Wa WT Z.

Cyclops macruroides denticulatus (Graeter) (*Leptocyclops lilljeborgi*). Female, 0·9 – 1·23 mm.
Norfolk.
Lake District: Br BT Hi LL Ly T W Z.

Cyclops macrurus (Sars). Female 1·1 – 1·2 mm. Colour yellow. Egg sacs very small, close to abdomen. Rather uncommon, but widely distributed.
Lake District: Bl C D Ea(?) Es G L R W Wa.

Sub-genus PARACYLOPS (PLATYCYCLOPS)

Cyclops fimbriatus (s. str.) (Fischer). Female, 0·86 – 0·9 mm. Egg sacs closely apposed to abdomen, with few large eggs. Adaptable species, able to live in running water or in depths of lakes (down to 200 m); in a thin film of water or subterranean.
Generally distributed throughout the British Isles.
Lake District: B Bl BT Es Lp Ly R W Z.

Cyclops fimbriatus poppei (Rehberg). Female, 0·86 – 0·9 mm.
Southern England.
Lake District: not recorded.

Cyclops affinis (Sars). Female, 0·72 mm. Egg sacs rather small, closely pressed to abdomen, and with few, rather large eggs. Found creeping among weeds and decayed stems.
 Probably generally distributed throughout the British Isles.
 Lake District: B Br Dv E M W Z.

Sub-genus ECTOCYCLOPS (PLATYCYCLOPS)

Cyclops phaleratus (Koch). Female, 0·9 – 1·0 mm. Egg sacs closely pressed to abdomen. Body light brown; legs, part of thorax and end of abdomen and antennules blue. Rather rare, found singly, or in small numbers.
 Widely distributed in the British Isles.
 Lake District: B Bl L M Z.

Sub-genus CYCLOPS

Cyclops strenuus (s. str.) (Fischer). Female, 1·66 – 2·35 mm. Robust. Ponds and ditches only, often in such as dry up in summer (except in Norfolk Broads). Active reproduction usually restricted to winter and early spring; disappearing or becoming rare in summer.
 Typical form of *C. strenuus* very common in small ponds throughout England; recorded from Wales and Ireland, rare in Scotland.
 Lake District: not certainly recorded.

Cyclops strenuus abyssorum Sars. Female, 1·20 – 1·47 mm. Monocyclic species, breeding in late summer and autumn; found in plankton of most lakes, and in small tarns at all elevations.
 This is the common form of *C. strenuus* throughout Scotland and in the Lake District.
 Lake District: B(?) Br BT Bu BW C Cr Ea G Go H M T U W Wa Z.

Cyclops scutifer Sars. Female, 1·2 – 1·4 mm. Spermatophore attached obliquely, projecting beyond margin of somite. Egg sac small, with few eggs.
 It has not been found in Britain, but it is not impossible that it has been overlooked.

Cyclops furcifer Claus (*C. lacunae*). Female, 1·44 – 2·10 mm. Egg sacs smaller than in *C. strenuus*, but with numerous eggs. Colour conspicuous orange, obscured in mature females by the dark brown ovaries. Rare. Found only in winter and spring in pools drying up in summer.
 Southern half of England.
 Lake District: not recorded.

Cyclops vicinus Uljanin. Female, 1·44 – 1·85 mm. Spermatophores attached obliquely, not projecting beyond the somite. Egg sac large, with many eggs. The commonest planktonic species of *Cyclops*, but also found occasionally in small ponds, generally associated with *Daphnia pulex*. Chiefly a summer form.
 Southern counties of England, Cheshire, Dundee.
 Lake District: not recorded.

Sub-genus ACANTHOCYCLOPS

Cyclops viridis (Jurine) (*C. vulgaris*, *Megacyclops viridis*). Female, 1·5 – 3·0 mm. Robust. Egg sacs large, divergent. Very common especially in the weedy margins of lakes or small pools. Occasionally taken on the bottom in deep lakes, but not in plankton. Breeds throughout the year.
 Generally distributed throughout the British Isles.
 Lake District: A Br BW C D E Ea Ee F G Go Gr Lp M Sm St U W Wa WT Z.

Cyclops gigas (s. str.) (Claus). Female, 2 – 3 mm. Egg sacs very large, closely adpressed to abdomen. Monocyclic, adults found only during winter.
 Wicken Fen and probably elsewhere in the British Isles.
 Lake District: not recorded.

Cyclops gigas latipes (Lowndes). Female 1·85 – 2·25 mm. Egg sacs large, divergent. Colour greyish, never green.
 Southern half of England; Ireland.
 Lake District: not recorded.

Cyclops vernalis (s. str.) (Fischer) (*C. lucidulus, C. robustus*). Female, 1·0 – 1·8 mm. Cuticle sometimes pitted over the abdominal somites or the whole body. Robust. Egg sacs rather large, not divergent, with 7–70 eggs. Colour yellowish to reddish. Very common in ditches and small pools, but never in lakes or other open waters. Breeds throughout the year, but is most common in spring and autumn.

Southern and eastern counties of England; Yorkshire.
Lake District: Lp W Z.

Cyclops vernalis americanus (Marsh). Female, 1·3 – 1·5 mm. Cuticle without pit markings. Colourless, except for blue pigment in the walls of the gut and mouth region, and small blue spots in rostrum, labrum, appendages and abdomen. Planktonic.

Southern half of England; Yorkshire.
Lake District: not recorded.

Cyclops venustus (Norman & Scott). Female, 1·0–1·22 mm. Body rather robust. Cuticular markings in the form of irregular ridges on the thorax which become broken up into scale-like elevations on the abdomen. Egg sacs closely apposed to body, with few eggs. Associated with *Sphagnum*.

Northern England, Devon, Cornwall, Isle of Man, Ireland.
Lake District: Z.

Cyclops sensitivus (Graeter & Chappuis). Female, 0·72 mm. Leg VI represented by a long seta and a short spine. Egg-sacs long, divergent, with many eggs. Colourless. Subterranean.

Ringwood, Hants.
Lake District: not recorded.

Cyclops bicuspidatus (s. str.) (Claus) (*C. pulchellus*). Female, 0·95–1·57 mm. Egg sacs large, generally divergent. Colour variable, commonly whitish or pink, often yellow. Cuticle of whole body including antennules and furcal ramus generally covered with minute pits. Commonly in shallow pools and ditches, particularly those containing much decaying vegetable matter, and evidently able to survive drying in mud; also in the profundal of some lakes.

Widely distributed in Britain.
Lake District: B(?) Bl Es G R WT Z.

ECOLOGY

Cyclops bicuspidatus lubbocki (Claus) (*C. b. odessanus*).
Female about 1 mm. Ditches, also brackish pools.
British Isles, mainly south and east England.
Lake District: not recorded.

Cyclops bicuspidatus thomasi (S. Forbes). Female, about 1 mm.
One uncertain record from Duddingston Loch, Edinburgh.
Lake District: not recorded.

Cyclops bisetosus (Rehberg). Female, 0·84–1·2 mm. Egg sacs large, closely pressed to abdomen. Cuticle generally with fine pits. Found generally in small ponds, often of a temporary nature; tolerates brackish water.
Widely distributed in the British Isles.
Lake District: Bl Lp W Z.

Cyclops crassicaudis (Sars). Female, 0·93–1·0 mm. Egg sacs large, slightly divergent. Colour milk white. Found only in very shallow water on moorland or under shade of trees.
Norfolk; Lough Derg, Ireland.
Lake District: not recorded.

Cyclops languidus (Sars). Female, 0·73–0·94 mm. Egg sacs large, divergent. Small pools, or very wet moss.
Scattered records throughout British Isles.
Lake District: Ee Lp W Z.

Cyclops languidoides (s. str.) (Lilljeborg). Female, 0·72 – 0·75 mm. This form has not been seen in this country.

Cyclops languidoides hiberniae (Gurney). Female, 0·84 – 0·94 mm.
Co. Dublin, Pembroke and Anglesey.
Lake District: not recorded.

Cyclops languidoides hypnicola (Gurney). Female, 0·46 – 0·53 mm. Submerged moss in shallow water, and subterranean.
Norfolk, Hants, Berkshire, Scotland.
Lake District: not recorded.

Cyclops languidoides eriophori (Gurney). Female, 0·59 – 0·7 mm. Among decaying leaves.
    Marsh at Ingham, Norfolk.
    Lake District: not recorded.

Cyclops nanus (Sars) (*C. diaphanus*). Female, 0·7 - 0·72 mm. A summer form, found in mossy pools.
    Generally distributed throughout Britain.
    Lake District: Ee W Z.

Sub-genus MICROCYCLOPS

Cyclops varicans (s. str.) (Sars). Female, 0·6 - 0·9 mm. Egg sacs large, slightly divergent. Colour greyish white. One of the rarest species. Summer form, usually found singly, or in small numbers.
    Generally distributed in the British Isles.
    Lake District: W.

Cyclops varicans rubellus (Lilljeborg). Female, 0·6 – 0·9 mm. Summer form. Flooded mossy marshes often with *C. nanus*.
    England, Isle of Man, Ireland.
    Lake District: B L.

Cyclops bicolor (Sars). Female, 0·6 – 0·7 mm. Egg sacs large, close to abdomen. Cephalothorax colourless or with blue spots, antennules and abdomen a rich golden yellow.
    Generally distributed in the British Isles.
    Lake District: B Ea G R W Z.

Cyclops minutus (Claus). Female, 0·83 – 1·06 mm. Egg sacs rather large, not divergent. Usually in temporary pools. Rare.
    England and Wales.
    Lake District: not recorded.

ECOLOGY

Cyclops gracilis (Lilljeborg). Female, 0·78 – 0·82 mm. Egg sacs small, with few eggs, closely pressed to abdomen. Colour rosy-red, antennule and ramus particularly richly coloured. Small ponds.
  Norfolk.
  Lake District: not recorded.

Cyclops unisetiger (Graeter). Female, 0·37 – 0·52 mm. Striking resemblance to a harpacticid copepod. Egg sacs not seen. Colourless. Wet peat or moss. Subterranean.
  Oxford, Norfolk, North Wales.
  Lake District: not recorded.

Cyclops demetiensis (Scourfield). Female, 0·5 mm. Eye red, surrounded by a tripartite mass of blackish granules. Leg VI, well developed (in the female as well as male) as 3 setae on each side of the genital somite. Colour white. Egg sacs not seen. Possibly subterranean.
  Tenby, in a trickle of water from a fissure in the rocks.
  Lake District: not recorded.

Sub-genus MESOCYCLOPS

Cyclops leuckarti (Claus) (*Mesocyclops obsoletus*). Female, 0·88 – 1·13 mm. Egg sacs large, divergent. Colour greyish, sometimes with darker bands. Small pools and also in lake plankton. Summer form. Overwinters in Windermere as a dormant fifth copepodid (Fryer & Smyly 1954).
  Common throughout central and southern England, but rarer in the north; Ireland.
  Lake District: B Bl C(?) D Es Hi L R W Z.

Cyclops hyalinus (Rehberg) (*Mesocyclops crassus*). Female, 0·8 – 1·0 mm. Egg sacs apposed to abdomen, with few eggs. Colour yellowish or brownish yellow, legs brown, the spine of exopods purplish brown, some blue colour at the base and ends of the ramus. Found only in summer, either in ditches with clear water and rich vegetation, or in plankton in Norfolk Broads.
  East and south-east England; Devon.
  Lake District: not recorded.

Cyclops dybowskii (Lande). Female, 0·67 – 0·85 mm. Egg sacs rather divergent. Colour generally brownish, the ramus and legs darker, and commonly tinged with violet.

Widely distributed in England; also found in Scotland.

Lake District: G D Ly T W Z.

REFERENCES

Fryer, G. (1953). Notes on certain freshwater crustaceans. *Naturalist, Hull*, 1953, No. 846, 101–9.

Fryer, G. & Smyly, W. J. P. (1954). Some remarks on the resting stages of some freshwater cyclopoid and harpacticoid copepods. *Ann. Mag. nat. Hist.* (12) **7**, 65–72.

Fryer, G. (1955). A faunistic and ecological survey of the freshwater Crustacea of the Huddersfield district of west Yorkshire. *Naturalist, Hull*, 1955, No. 854, 101–26.

Galliford, A. L. (1954). Notes on the freshwater organisms of Lundy, with especial reference to the Crustacea and Rotifera. *Rep. Lundy Fld Soc.* **7** (1953), 29–35.

Galliford, A. L. & Williams, E. G. (1951). Microscopic organisms of some brackish pools at Leasowe, Wirral, Cheshire. *NWest. Nat.* **23**, 39–62.

Grainger, J. N. R. (1952). The *Diaptomus* of some lakes in S.W. Ireland. *Proc. R. Ir. Acad.* **54** (B), 217–24.

Gurney, R. (1931-33). *British Freshwater Copepoda.* Vols. I–III. London (Ray Society).

Kiefer, F. (1929). Crustacea Copepoda. 2. Cyclopoida Gnathostoma. *Tierreich*, lief 53.

Sars, G. O. (1901-03). *An account of the Crustacea of Norway with short descriptions and figures of all the species.* Vol. IV. Copepoda, Calanoida. Bergen.

Sars, G. O. (1913-18). *Crustacea of Norway.* Vol. VI. Copepoda, Cyclopoida. Bergen.

Slack, F. E. (1956). A key to the British species of the genus *Cyclops* O. F. Müller. *Glasg. Nat.* **17**, 250–6.

INDEX

Names in parentheses are synonyms.
Bold figures indicate illustrations.
Keys and illustrations are on pp. 6–41, ecological notes on pp. 42–54.

Acanthocyclops (sub-genus) 40–41, 49
Acartia 9, 11
   bifilosa **10**, 11, 45
   clausi **10**, 11, 44
   discaudata **10**, 11, 45
   (*ensifera*) 44
Calanoida 7 ff., 43–5
Canthocamptus pygmaeus [not included in key] **6**
Centropages hamatus **8**, 9, 43
Cyclopina norvegica **32**, **36**, 37, **38**, 45
Cyclopoida 25 ff., 45–54
Cyclops (sub-genus) 40, 47
Cyclops affinis 39, 48
   agilis **34**, 39, 46
   agilis speratus 39, 47
   albidus **26**, **36**, 37, 39, 46
   bicolor 41, 52
   bicuspidatus 40, 50
   bicuspidatus lubbocki 40, 51
   (*bicuspidatus odessanus*) 51
   bicuspidatus thomasi 40, 51
   bisetosus 40, 51
   crassicaudis 40, 51
   demetiensis **32**, 41, 53
   (*diaphanus*) 52
   distinctus 39, 46
   dybowskii 41, 54
   fimbriatus **36**, 39, 47
   fimbriatus poppei **36**, 39, 47
   furcifer 40, 49
   fuscus **26**, **30**, 39, 45
   gigas 40, 49
   gigas latipes 40, 49
   gracilis 41, 53
   hyalinus **32**, 41, 53
   (*lacunae*) 49
   languidoides 41, 51
   languidoides eriophori 41, 52
   languidoides hiberniae 41, 51
   languidoides hypnicola 41, 51
   languidus **24**, **28**, 40, 51
   leuckarti **26**, **36**, 37, 41, 53
   (*lucidulus*) 50
   macruroides **26**, 39, 47
   macruroides denticulatus **26**, 39, 47
   macrurus 39, 47
   minutus **30**, **34**, 37, 41, 52
   nanus **32**, 41, 52
   phaleratus **32**, **36**, 39, 48
   prasinus **26**, **32**, 39, 46
   (*pulchellus*) 50
   (*robustus*) 50
   scutifer **24**, 37, 40, 48
   sensitivus **32**, 40, 50
   strenuus 40, 48
   strenuus abyssorum **26**, **32**, 40, 48

   unisetiger **30**, **32**, **34**, 37, 41, 53
   varicans **32**, 41, 52
   varicans rubellus **32**, 41, 52
   venustus 40, 50
   vernalis **32**, **34**, 40, 50
   vernalis americanus 40, 50
   vicinus **30**, **34**, 37, 40, 49
   viridis 40, 49
   (*vulgaris*) 49
Diaptomus 9, 15 ff., 22–3
   (*bacillifer*) 44
   castor **14**, **16**, **18**, **20**, 22–3, 43
   gracilis **14**, **16**, **18**, **20**, 22–3, 43
   (*hircus*) 44
   laciniatus **16**, **20**, 22–3, 43
   laticeps **16**, **18**, **20**, 22–3, 44
   (*serricornis*) 44
   vulgaris **16**, **18**, **20**, 22–3, 44
   wierzejskii **14**, **16**, **18**, **20**, 22–3, 44
(*Dias longiremis*) 44
Ectocyclops (Platycyclops) (sub-genus) 39, 48
Eucyclops (Leptocyclops) (sub-genus) 39, 46
(*Eucyclops serrulatus*) 46
Eurytemora 9, 13
   affinis **12**, 13, 44
   americana **12**, 13, 44
   (*clausi*) 44
   (*hirundoides*) 44
   (*lacustris*) [not included in key] 13
   (*Thompsoni*) 44
   velox **12**, 13, 44
Halicyclops 42
   (*aequoreus*) 45
   (*aequoreus propinquus*) 45
   christianensis **24**, **32**, **38**, 45
   (*magniceps*) 45
   neglectus **38**, 45
Harpacticoida [not included in key] 6, 7
(*Leptocyclops agilis*) 46
   (*lilljeborgi*) 47
Limnocalanus macrurus 9, **10**, 43
Macrocyclops (sub-genus) 39, 45
(*Megacyclops viridis*) 49
Mesocyclops (sub-genus) 41, 53
(*Mesocyclops crassus*) 53
   (*obsoletus*) 53
Microcyclops (sub-genus) 41, 52
(*Pachycyclops annulicornis*) 46
   (*bistriatus*) 46
   (*signatus*) 45
Paracyclops (Platycyclops) (sub-genus) 39, 47
Tropocyclops (sub-genus) 39, 46